Water

Developed at
The Lawrence Hall of Science,
University of California, Berkeley
Published and distributed by
Delta Education,
a member of the School Specialty Family

© 2012 by The Regents of the University of California. All rights reserved. No part of this book may be reproduced or transmitted in any form or by any means, electronic or mechanical, including photocopying or recording, or by any information storage and retrieval system, without permission in writing from the publisher.

1325247
978-1-60902-040-8
Printing 3 — 7/2013
Quad/Graphics, Versailles, KY

Table of Contents

Investigation 1: Water Observations
A Report from the Blue Planet 3
Surface Tension 6
Which Way Does It Go? 8

Investigation 2: Hot Water, Cold Water
Water: Hot and Cold 10
Ice Is Everywhere 13

Investigation 3: Water Vapor
Drying Up 19
Surface-Area Experiment 21
The Water Cycle 23

Investigation 4: Waterworks
Water: A Vital Resource 26
Natural Resources 31
The Power of Water 35
Ellen Swallow Richards: An Early Ecologist 38
Solar Disinfection System 42

References
Science Safety Rules 44
Glossary 45
Index 48

A Report from the Blue Planet

TO: Chief of Science, Home Planet
FROM: Interplanetary Science Office, Fleet 2012

Greetings from the blue planet mentioned in my last report. We have been exploring the planet as directed. Now we know why it looks blue from space. Almost three-quarters of the planet's surface is covered by **water**! In all our planet explorations, this is the first water planet we have discovered. Here's what we have learned so far.

Ninety-seven percent of the planet's water is in its huge ocean of **salt water**. Our first view of the blue planet was almost all ocean. When we flew around to the other side, we saw that there is dry land, too.

A view of the Atlantic Ocean

A view of the Pacific Ocean

The rest of the planet's water is **fresh water**. That means only 3 percent of the water is free of salt. And about two-thirds of the fresh water is **solid ice**. That leaves just 1 percent of the planet's water as **liquid** fresh water.

Ocean

Liquid fresh water is found in many places. A lot of the water is underground. The rest of the fresh water is on the planet's solid surface. We see it in lakes and rivers. All the plants and animals on the blue planet need water to stay alive. The people living there use water in many ways. They use it for cooking, washing, drinking, and growing food.

We have observed water in three states on the planet. It is the only material found naturally on this planet in all three states. Water can be solid ice, liquid water, and a **gas** called **water vapor**.

Water vapor is in the air. There is more water in the air than in all the rivers on the planet. We will find out more about water vapor for our next report.

As you can see, water is an amazing material. It is in the ocean, in lakes and rivers, in the ground, and in the air. It is everywhere.

Fresh Water 3%
Ice 2%
Liquid 1%

Glacier

Lake

River

Surface Tension

Have you ever seen an insect walk on water? If you have, you may have wondered, how can it do that? The answer is **surface tension**. Water is made of tiny particles. The particles are naturally attracted to one another. At the surface, where water meets the air, the attraction between particles is very strong. The strong attraction at the surface of the water is surface tension. Insects like water striders can walk on water because bristles on their feet keep them from breaking through the surface tension.

You can see how surface tension works. Fill a glass to the top with water. Keep adding more water a little at a time. If you are careful, you can "overfill" the glass. The water will form a dome above the top of the glass, but it won't spill out. Why? Surface tension.

What happens when water falls through air? Water forms drops. Drops are small **volumes** of water with surface all around. The skinlike surface tension pulls all around the outside of the drop. The pulling results in a perfect sphere. Next time you are in the shower, look closely at the falling water. What do you see?

Without surface tension, rain falling from clouds might fall in sheets or strings. Without surface tension, water landing on a window, a car, or a leaf would spread out into a thin film. But the water doesn't make a thin film. It forms dome-shaped bits of water called **beads**. Why do you think water forms beads when it falls on a waterproof surface?

Remember surface tension the next time you watch water striders zip across the water. The little insects glide over water as if they were skating on ice because of surface tension.

Water forms beads.

7

Which Way Does It Go?

Go outside during a rainstorm and look around. What happens to the rainwater? Some of it **soaks** down into the **soil**. Some of it flows across the ground, sidewalks, and paved surfaces. Water always seems to be on the move. Why is that? Let's follow a few drops of water that are on the move.

Look on top of the mountain. There is still some snow high up around the peaks. Drops of melted snow flow down the sides of mountains and into brooks. Brooks join to form streams, and streams tumble over cliffs as waterfalls.

Streams flow into rivers. Drops of water in rivers slow down when the river is dammed. But they don't stop. When water drops pass over the dam, they flow to the ocean. When water gets to the ocean, it finally stops moving. Or does it? There may be more to the story of a water drop.

Look again at the pictures. Follow the water drops from the mountain peak to the ocean. Which way does water go? Water always flows in the same direction. Water always moves down.

Water is **matter**. Like all matter, water is pulled downward by **gravity**. That's why brooks flow from mountain peaks to forest meadows. That's why meadow streams flow into river valleys. That's why rivers flow down to the ocean.

The next time it rains, watch the water flowing across the ground. Which way is it going?

Water: Hot and Cold

When things get hot, something interesting happens. They get bigger. Usually you can't see that the hot material is bigger. The change is small. But one place you can see that hot material is bigger is in a bulb **thermometer**.

A bulb thermometer is a small container of liquid attached to a thin tube. The small container is the bulb. The thin tube is the stem. When the bulb gets hot, the liquid **expands** (gets larger). Liquid pushes farther up the stem. When the bulb gets cold, the liquid **contracts** (gets smaller). Liquid pulls back into the bulb.

How does that happen? It happens at a level that is invisible to our eyes.

This is what scientists have figured out. Water is made of tiny particles that are much too small to see. The particles are moving around all the time. They move faster when the water is hot and slower when the water is cold.

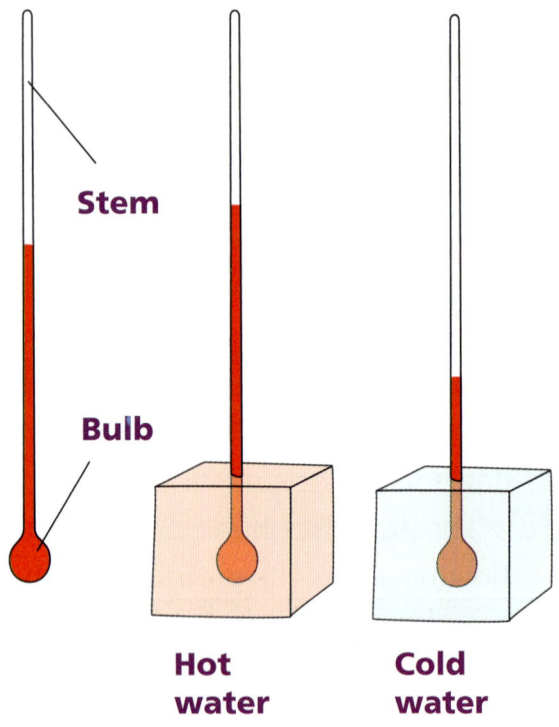

Think about a pan of water. All the water particles bang into one another all the time. That keeps a little space between the particles. When the water is hot, the particles bang into one another harder. Harder banging pushes the particles a little farther apart. When the particles are farther apart, the volume of water in the pan increases. Increased volume is **expansion**.

Now can you explain what happens to the liquid in a bulb thermometer?

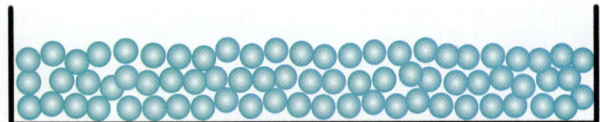

Particles of cold water in a pan

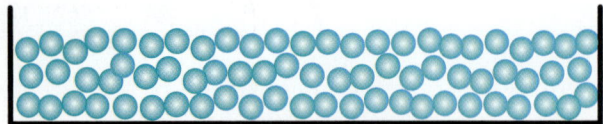

Particles of hot water in a pan

Float and Sink

Imagine that you are having sunflower seeds for a snack and that they spill. The seeds fall onto gravel where they are hard to see. How could you separate this **mixture** of seeds and gravel? Just scoop up the seeds and gravel and drop them into a bowl of water. The pieces of rock (gravel) will **sink**. The sunflower seeds will **float**.

Why do the seeds float and the bits of rock sink? Some might say it is because rocks are heavier than sunflower seeds. But that wouldn't be true.

Think about this. A piece of gravel on one side of a balance and a seed on the other side have the same **mass**. Each has a mass of exactly 0.1 gram (g).

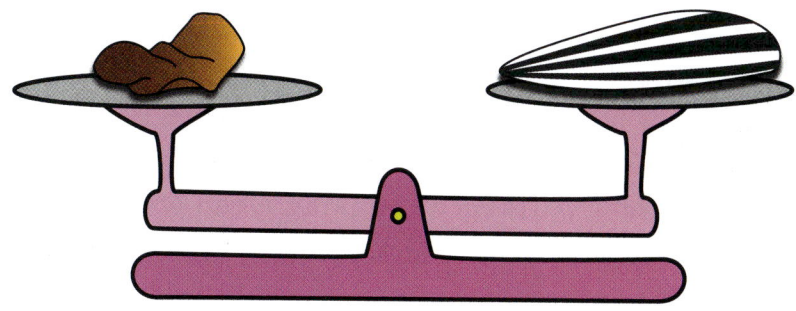

A small piece of gravel and a large sunflower seed have the same mass.

11

If we drop these two objects in water, the seed will still float, and the rock will still sink. Why? Because the volumes are different. The two objects have the same mass, but the mass is more concentrated in the piece of rock. The rock is **more dense** than the seed.

Density is the amount of mass compared to the volume. Imagine that we can scrunch both objects into perfect spheres. The mass will still be the same, but now we can compare the volumes.

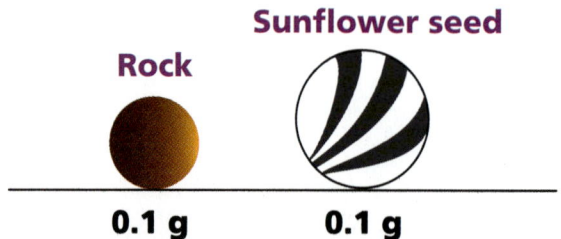

The rock has the same mass as the sunflower seed, but in a smaller volume, so the rock is more dense. But why does the rock sink and the seed float? Look at the same mass of water.

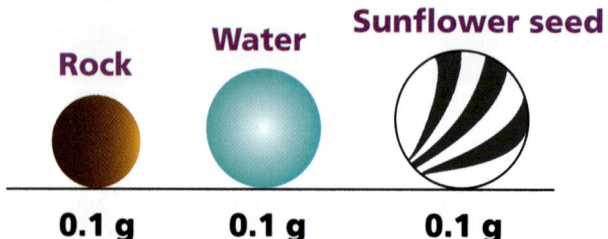

Compare the rock and sunflower-seed spheres to an equal mass of water. The volume of the water is larger than the volume of the rock, but smaller than the volume of the sunflower seed. The rock is more dense than the water, so it sinks. The sunflower seed is less dense than the water, so it floats.

Water Density

Why does warm water form a layer on top of room-temperature water? Water particles move faster when water gets hot. Particles push one another farther apart. The water expands. When water expands, the mass stays the same, but the volume increases. What happens to the density?

Look at the glass of layered water. Which is the hot water? Which is the cold water? How do you know?

Ice Is Everywhere

You probably know where to go to get some ice. Indoors you find a refrigerator and look in the freezer. Outdoors is a different story. If it's winter, and you live between Montana and Maine along the border with Canada, ice is everywhere. Every pond, creek, and bucket of water is frozen. In the warmer parts of the country, and during the summer, finding ice outdoors can be a challenge.

Some places are cold all year long. Alaska, Canada, Greenland, Iceland, Scandinavia, and Siberia have ice year-round. Antarctica, which covers the South Pole, is the iciest continent. More than 95 percent of its land lies under thick ice. In some places, the ice is 4,300 meters (m) thick. In the winter, frozen sea water around Antarctica doubles the continent's size!

If you live in snow country, you know what to expect. Usually starting in December, heavy snow falls, covering everything under a white blanket. During a heavy snow year, the snow may stay on the ground until March or April. Then it **melts**.

Polar bears on the ice in the Arctic

Ice off the coast of Greenland

13

Glaciers

What if the winter snow didn't melt during the summer? In some of the colder regions around the world, more snow falls than can melt in the summer. Snow piles up and up. The layers of snow at the bottom get compressed and turn into pure ice. When the ice is about 18 m thick, it begins to move. Moving ice is a **glacier**. Glaciers are "rivers" of ice that gravity pulls downhill.

Scientists can keep track of how fast glaciers move. An average glacier advances less than 1 m each day. A glacier in Greenland holds the speed record. Jakobshavn Glacier is speeding along at more than 35 m per day.

Glaciers now cover about 10 percent of Earth's land. They are found in all of the world's major mountain ranges. All the glaciers in the world store about 65 percent of the world's fresh water. If they all melted, sea level would rise about 79 m.

An Alaskan glacier

A glacier ends at the sea.

Icebergs

Icebergs are "islands" of ice drifting in the ocean. Icebergs are frozen fresh water, not salt water. Where does all the frozen fresh water come from?

Icebergs come from glaciers. When a glacier moving down a valley reaches the sea, pieces at the end break off. These chunks of ice may be as small as cars or as big as mountains. The largest iceberg ever measured was 320 kilometers (km) long!

We see only a small part of an iceberg. Seven-eighths is hidden beneath the water's surface. Icebergs in the North Atlantic can last up to 2 years before melting. Larger icebergs in the Antarctic may last 10 years.

Someday icebergs may be a useful source of fresh water. Ice could be harvested and melted. Whole icebergs might even be towed to countries needing fresh water!

Icebergs form when pieces of ice break away from the face of a glacier.

A large iceberg might extend a kilometer below the surface of the sea.

Ice History

Before refrigeration, people used ice to keep food cool. Ice was harvested from frozen lakes and rivers in winter. People waited until the ice was at least 60 centimeters (cm) thick. Then it was strong enough to hold the ice workers.

Horses were used to plow a frozen lake to clear away the snow. Then the horses pulled a special tool that scratched lines in the ice. Workers cut along the lines with sharp saws. They used poles to push large sheets of ice to icehouses. There they cut the sheets into smaller blocks.

Icehouses looked like barns. Inside, the ice workers carefully stacked and stored the ice blocks. They spread straw or sawdust over and around each block to keep it from melting and sticking to other ice blocks.

Ice workers mark lines on the frozen lake.

Large sheets of ice are pushed to the icehouse.

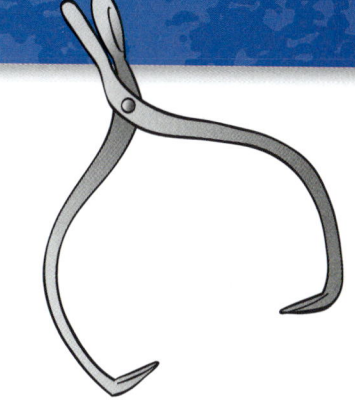

Most homes had an icebox. Throughout the year, horse-drawn trucks carried blocks of ice to homes in towns and cities. The iceman used ice tongs to handle the heavy blocks. Ice blocks could range from 11 to 22 kilograms (kg) each. The iceman then used an ice pick to fit the block of ice inside the icebox.

Children loved to see the ice truck on hot summer days. They crowded around when the iceman pushed the door open. There was sure to be a sliver of ice for each of them. What a treat on a steamy day!

Why Pipes Burst

Pipes supply water to houses, schools, and other buildings. They are made of strong materials, like plastic, copper, cast iron, and steel. But sometimes pipes burst! Do you know why?

Water, like all other materials, contracts as it cools. But once the temperature gets down to 4 degrees Celsius (°C), an amazing thing happens. As water continues to get colder, it starts to expand. Between 4°C and 0°C (the temperature at which water **freezes**), water expands. That means the ice needs more space as it changes from liquid to solid. If liquid water completely fills a container, it will break the container when it freezes. Ice needs room to expand, or it will break its container.

How can you prevent pipes from breaking in really cold weather? If the pipes are not used during the winter, drain the water out. If that is not possible, make sure pipes are well insulated. Pipes can be wrapped with insulation, or they can be buried deep underground.

Another solution is to leave the tap open just a little bit. Then the expanding water can push out the end of the pipe. The water might freeze in the pipe, but the pipe will not break. If the tap is closed, the water is trapped. Then something will break when the water freezes.

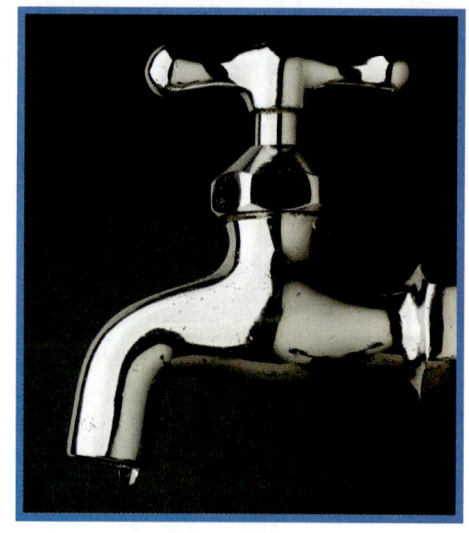

Remember those icebergs in the northern and southern seas? Why do they float? There is a connection between floating icebergs and breaking water pipes. Do you know what it is?

Why do pipes break? Because water expands as it turns to ice. The amount (mass) of water doesn't change, only its volume. If the mass stays the same but the volume increases, the density of ice changes. If ice is less dense than liquid water, what will happen when you put ice in water? It floats.

Drying Up

You know when something is wet. It is covered with water, or it has soaked up a lot of water. When it rains, everything outside gets wet. When you go swimming, you and your swimsuit get wet. Clothes are wet when they come out of the washer. A dog is wet after a bath.

But things don't stay wet forever. Things get dry, often by themselves. An hour or two after the rain stops, porches, sidewalks, and plants are dry. After a break from swimming to eat lunch, you and your swimsuit are dry. After a few hours on the clothesline, clothes are dry. A dog is dry and fluffy after a short time. Where does the water go?

You can't see water vapor in the sky.

The water **evaporates**. When water evaporates, it changes from water in its liquid form to water in its gas form. The gas form of water is called water vapor. The water vapor leaves the wet object and goes into the air. As the water evaporates, the wet object gets dry.

What happens when you put a wet object in a sealed container? It stays wet. If you put your wet swimsuit in a plastic bag, it's still wet when you take it out of the bag. Why? A little bit of the water in your suit evaporates, but it can't escape into the air. The water vapor has no place to go, so your suit is still wet when you get home.

Have you ever seen water vapor in the air? No, water vapor is invisible. When water changes into vapor, it changes into individual water particles. Water particles are too small to see with your eyes. The water particles move into the air among the nitrogen and oxygen particles. When water becomes part of the air, it is no longer liquid water. It is a gas called water vapor.

Surface-Area Experiment

Julie and Art want to find out how **surface area** affects **evaporation**. They decide to do an experiment. They have some plastic boxes to put water in, some graph paper, and a set of measuring tools. They are ready to start.

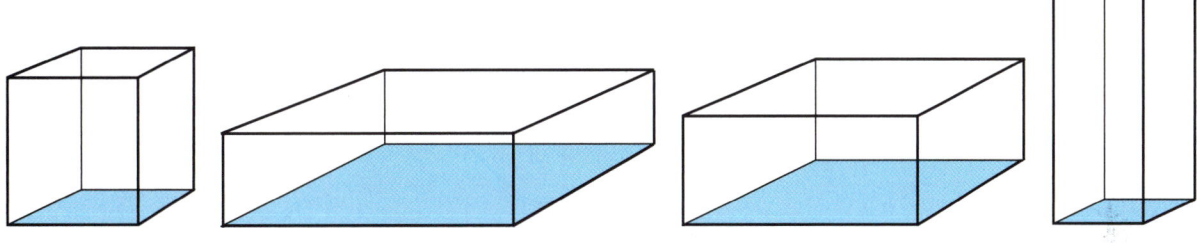

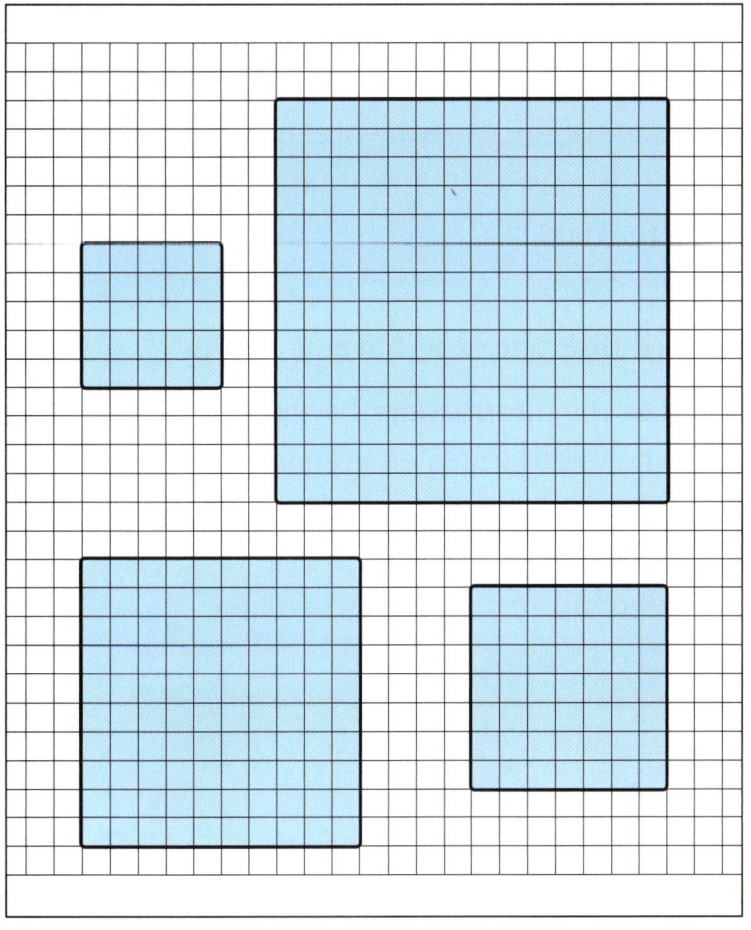

Julie has an idea for measuring the surface area of each box. She traces around each box on graph paper. She uses a meter tape to measure the distance between the lines on the graph paper. The lines are 1 centimeter (cm) apart.

21

The two students number the boxes. The box with the smallest surface area is number 1. The box with the biggest surface area is number 4. Then they measure 50 milliliters (mL) of water into each box. They place the four boxes on the counter by a window.

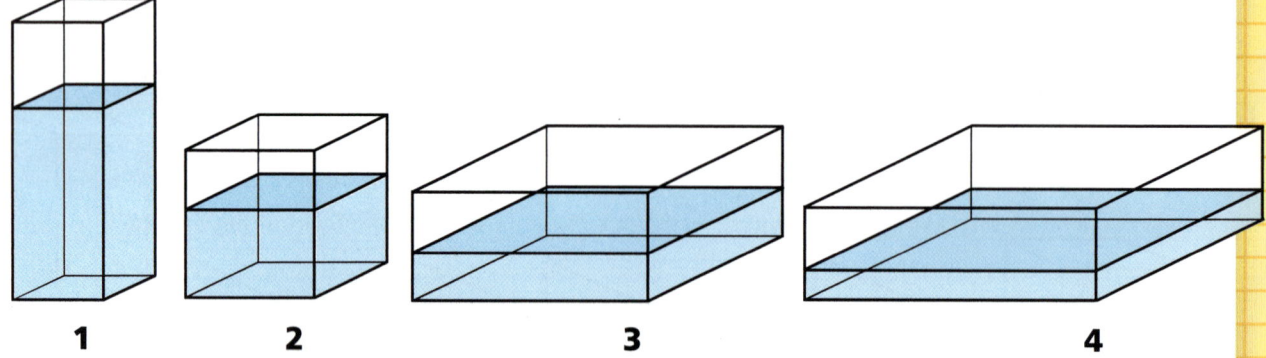

One week later, Julie and Art measure the amount of water in each box. Box 1 has 46 mL, box 2 has 42 mL, box 3 has 34 mL, and box 4 has 18 mL.

Art thinks about the results. It seems that the surface area of the water in the boxes has an effect on the evaporation. But he isn't sure. Julie suggests organizing the results of the experiment. The students decide to do the following.

- Make a T-table to display the data.
- Make a graph of the data.
- Describe what they learned from the experiment.

Can you help Julie and Art? Use the information they gathered to write a report about the effect of surface area on evaporation. Be sure to include the three kinds of information listed above.

Thinking about the Experiment

What additional information would be useful to better understand how surface area affects evaporation?

The Water Cycle

The amount of water on Earth is fairly constant. We use it over and over again. It's recycled. Water you swallow today may have once rained into the Grand Canyon. It may have washed one of Abraham Lincoln's shirts. It might even have been a dinosaur's drink!

The recycling of water is the **water cycle**. Energy from the Sun evaporates water. Most evaporation occurs at the ocean's surface. Water also evaporates from lakes, rivers, puddles, and clothes hung on a line. Evaporation changes liquid water into water vapor.

Water vapor is a gas. Water vapor enters the air. It becomes part of the air. You don't see water vapor in the moist air because it's invisible. Air moves freely from place to place as wind. Water vapor in air is free to move all over the world.

Do you remember what happens to water vapor when it contacts a cold surface? It **condenses**. The change of state from gas to liquid is **condensation**. You observed condensation when water vapor formed droplets on the outside of a cup of ice water. You may have seen condensation on cool windows and mirrors, too.

Moist air that condenses outdoors at ground level is called **dew**. You can see dew in the morning as tiny droplets of liquid water on surfaces such as blades of grass and spider webs.

Warm air carries water vapor higher into the sky. As air rises, it cools. When water vapor touches cold surfaces in the air, like bits of dust, water vapor condenses. At first, each droplet is too small to see. But droplets bump into each other and join together. The droplets grow larger and larger. When millions and millions of these clumps of water droplets come together, they form clouds.

If it is cold, the water droplets may form ice crystals. When the water droplets or ice crystals get too heavy to stay in the air, they fall to Earth as **precipitation**. Rain, snow, sleet, and hail are forms of precipitation. The ocean covers more than 70 percent of Earth. Most precipitation falls in the ocean, where the Sun will once again evaporate the water. Energy from the Sun drives the water cycle.

On land, raindrops are taken in by plant roots. They escape back into the air through holes in plant leaves. Other raindrops sink into the ground. Ground water flows through sand and pebbles and might seep slowly through clay. Ground water moves slowly. It might take 100 years for some water drops to return to Earth's surface in a spring.

Raindrops also land on rocks, roofs, and roads. Sunshine evaporates more than half of the raindrops that fall, returning water vapor to the air. Some raindrops run along the ground. Those drops might flow into a stream, a river, a **reservoir**, and finally, after a long journey, back to the ocean.

Water drops that fall on ice sheets and glaciers take the longest time to return to water vapor. Water may stay frozen for hundreds or even thousands of years! But all water is part of the water cycle that recycles the same supply of water over and over. Water evaporates into water vapor. Vapor condenses into liquid water. Liquid water falls back to Earth. It happens every day, every week, every year, forever. It's a never-ending cycle that supports life on Earth.

Water Cycle

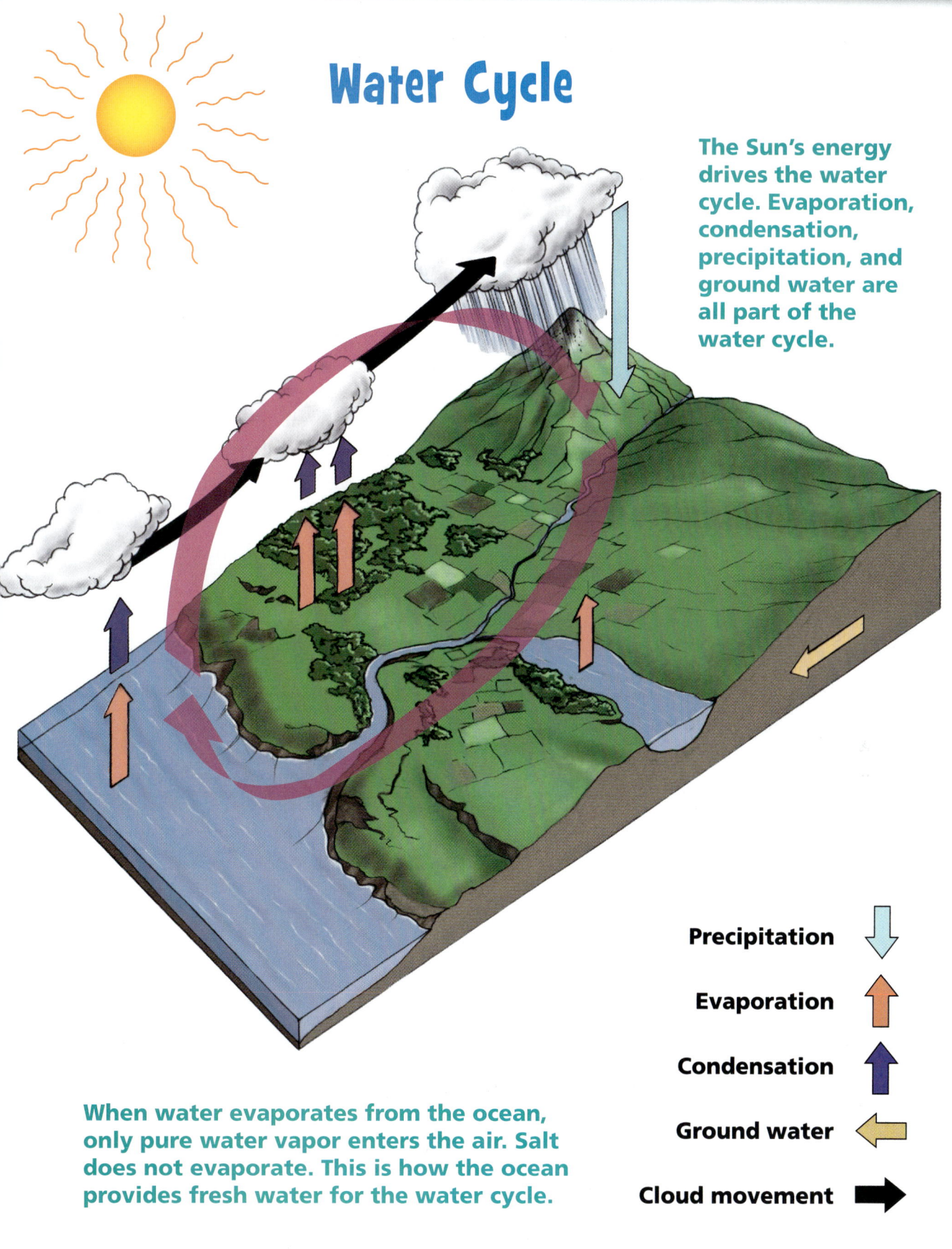

The Sun's energy drives the water cycle. Evaporation, condensation, precipitation, and ground water are all part of the water cycle.

When water evaporates from the ocean, only pure water vapor enters the air. Salt does not evaporate. This is how the ocean provides fresh water for the water cycle.

Precipitation
Evaporation
Condensation
Ground water
Cloud movement

WATER: A Vital Resource
by Keira, David, Tamiko, and Jorge

Our team's assignment was to learn about our water supply. Earth seems to have plenty of water. But 97 percent of that water is salt water. Another 2 percent of the world's water is frozen. That leaves just 1 percent as fresh water. The good news is that 1 percent should be enough for everyone. The bad news is that it's not spread equally around the world. Some places have a lot of fresh water, but others do not.

Tamiko brought some information to our first meeting. She said we use 35 times more water today than people did 300 years ago. The human population has grown, and so have the ways we use water.

Jorge looked surprised and asked, "Will we ever run out of water?" Keira was sure the answer was no. She reminded us about the water cycle. She said, "The amount of water on Earth doesn't get used up. It gets recycled."

But David wondered if the amount of water we need will grow larger than the amount we have. That made all of us realize how important it is to take care of the water we have. Tamiko summed it up this way. "Water is one of the most valuable **natural resources** on Earth! We have to take care of it. If we make our water too dirty to use, or if we use our water faster than it is replaced, we will be in a lot of trouble."

At the end of the meeting, each of us chose something to investigate about our water supply. We agreed we had a lot to learn.

Water is a renewable resource, but it is not unlimited.

Water Coming into Our Homes
by Keira

My community takes water from Lake Charles. In other places, water comes from rivers or underground **aquifers**. Aquifers contain water that has soaked into the ground and is stored in layers of rock. Water is usually treated before it reaches our faucets. Water-treatment plants filter and treat the water, making it clear and safe for people to use.

First, water is screened to remove fish, leaves, and large objects such as logs or trash. Next, a machine called a flash mixer stirs the water with chemicals. Four chemicals are commonly mixed with water. They are lime, carbon, chlorine, and alum. Lime softens the water. Carbon **absorbs** materials that smell bad. Chlorine kills bacteria. Alum makes particles of clay clump together.

The mixed water goes to a settling tank. Clay clumps, silt, and other particles drift to the bottom. From there, the water passes through sand, gravel, and charcoal filters. A chemist at the water-treatment plant tests the water every day. This is to make sure that all harmful bacteria are killed. Purified water is pumped into tanks and towers. It reaches our homes through underground pipes.

Water Leaving Our Homes
by David

Water leaves our homes. It runs down sink and bathtub drains and out of washing machines. It is flushed down toilets. Waste water must be treated before it returns to the environment. In some communities, waste water goes to sewage-treatment plants. In other places, waste water enters local septic systems.

Septic tanks are usually made of concrete or metal. They are buried outside houses. Waste water separates inside a septic tank. Heavy materials sink to the bottom and form sludge. Lighter materials like fats and grease rise and form scum. Bacteria break down solids in the tank. The liquid in the middle flows through pipes into gravel-filled trenches. The liquid in the trenches is purified as it seeps through the gravel and soil.

Sewage-treatment plants screen waste water to remove solids. Bacteria break down other materials. Chemicals are used to rid the water of impurities. Treated water then discharges into streams, lakes, or the ocean.

City Runoff
by Tamiko

Rain does not always evaporate or soak into the ground. Sometimes it becomes **runoff**. Runoff flows over land and streets, and then into storm drains. Storm drains often empty right into bays, lakes, and streams.

Some people don't know that storm drains connect to local water systems. They sometimes pour pet waste, oil, paint, and other hazardous materials into storm drains. The untreated water harms the water supply.

In some cities, science clubs or environmental groups paint pictures of fish on the sidewalks near storm drains. The fish remind people that whatever goes into the storm drain will enter the water supply.

Water Conservation
by Jorge

Conserving water is an important part of protecting it. Because of conservation, water use in the United States has dropped since 1980. Here are a few things we can do to save water.

- Turn off the tap when you brush your teeth. Don't run water while washing dishes. Shut the shower off while you soap up.
- Take shorter showers and use a low-flow showerhead.
- Install low-flow aerators on all your faucets. An aerator mixes air with the water. You use less water when air is mixed in. The flow will still seem strong.
- Fix leaks in pipes, faucets, and toilets. Dripping faucets can waste about 7,500 liters (L) of water each year. Leaky toilets can waste as much as 750 L each day.
- Use less water to flush your toilet. Install a low-flow toilet, or put a water-filled plastic container in the tank if you have an older toilet.
- Use a broom, not a hose, to clean driveways and sidewalks.
- Water lawns and other outdoor plants in the morning. (Water evaporates faster in the middle of the day.) Don't water on a windy day.
- Put mulch around plants to reduce evaporation.

Thinking about Water

1. What is the source of your local water?
2. How is water purified in your community?
3. What are the issues about water in your community?

Natural Resources

Some people call it "dirt." Others call it "earth" or "the ground." What they are talking about is soil. Soil is the layer on top of the land. Soil is what you dig up with a shovel. You can stir soil with water to make mud or turn it over with a plow.

The soil in your schoolyard is different from the soil in a field. The soil in a field is different from the soil in a desert. In fact, soils are different just about every place you look. But in some ways, soils all over the world are the same.

All soils have two basic ingredients: rock and **humus**. The rock part of the soil comes in a variety of sizes, including gravel, sand, silt, and clay. Particles of gravel are rocks the size of rice and peas. Sand particles are smaller rocks. Silt particles are so small it's impossible to see just one. Clay particles are smallest of all.

Humus is black **decomposing organic matter**. It comes from the dead and discarded parts of plants and animals.

Soil in a field prepared for planting

Soil in the Mojave Desert

But soils have different **properties**. They differ in **texture** and color. Soils also differ in their ability to **retain** water and to support plant growth. The texture of soil depends on the amount and size of the rock particles. Soils with a lot of sand and gravel feel gritty. They fall apart easily when you make a mud ball. Soils with a lot of silt and clay feel smooth and slippery. They make excellent mud balls.

Mud balls made out of sandy soil and clay soil

Soil color depends on the color of the sand, silt, and clay particles, and the amount of humus. Texture and humus determine the amount of water that a soil can retain. If the soil has a lot of sand and gravel particles, water will flow through the spaces between the particles. Very little water will stay in the soil. Soil with smaller particles and a lot of humus will retain more water. The water gets trapped in the smaller spaces between particles. And water is absorbed by the organic humus particles.

Soil as a Natural Resource

Materials that people get from the natural environment are natural resources. Most of the food that people eat comes from plants or from animals that eat plants. Plants grow in soil. The soil is a natural resource that people depend on for survival.

Soil is a **renewable resource**. That means that natural processes make new soil, but it happens very slowly. Soil must be used wisely. If the soil resource is overused, it will lose its ability to support the growth of plants. Farmers need to renew the soil nutrients to make sure plenty of food crops will be available for people.

Other Natural Resources

People rely on many other natural resources. There are renewable resources and **nonrenewable resources**.

Renewable resources are replaced as we use them. We have investigated one important natural resource, water. We know that water is renewed all the time by the water cycle. Plants and animals are also renewable resources. New plants and animals are growing all the time. We use them for food and shelter. Wood is another example of a renewable plant resource.

When nonrenewable resources are used up, they are gone. People use a lot of nonrenewable natural resources as **energy sources**. Coal, petroleum, and natural gas are energy sources. These **fossil fuels** are the remains of plants and animals that lived millions of years ago. When Earth's fossil fuels are used up, they will be gone forever. No new fossil fuels are "growing" at this time. The length of time that we have fossil fuels can be extended by conservation. People can conserve fossil fuels by using more energy-saving products, like high-mileage cars and better insulated homes.

Lumber is a renewable resource.

Petroleum and coal are nonrenewable resources.

33

Some natural resources are **perpetual renewable resources**. Perpetual means they are available all the time whether we use them or not. Examples of perpetual renewable resources are solar energy, wind power, geothermal energy, and tides. The most important energy sources for the future will be based on energy from the Sun.

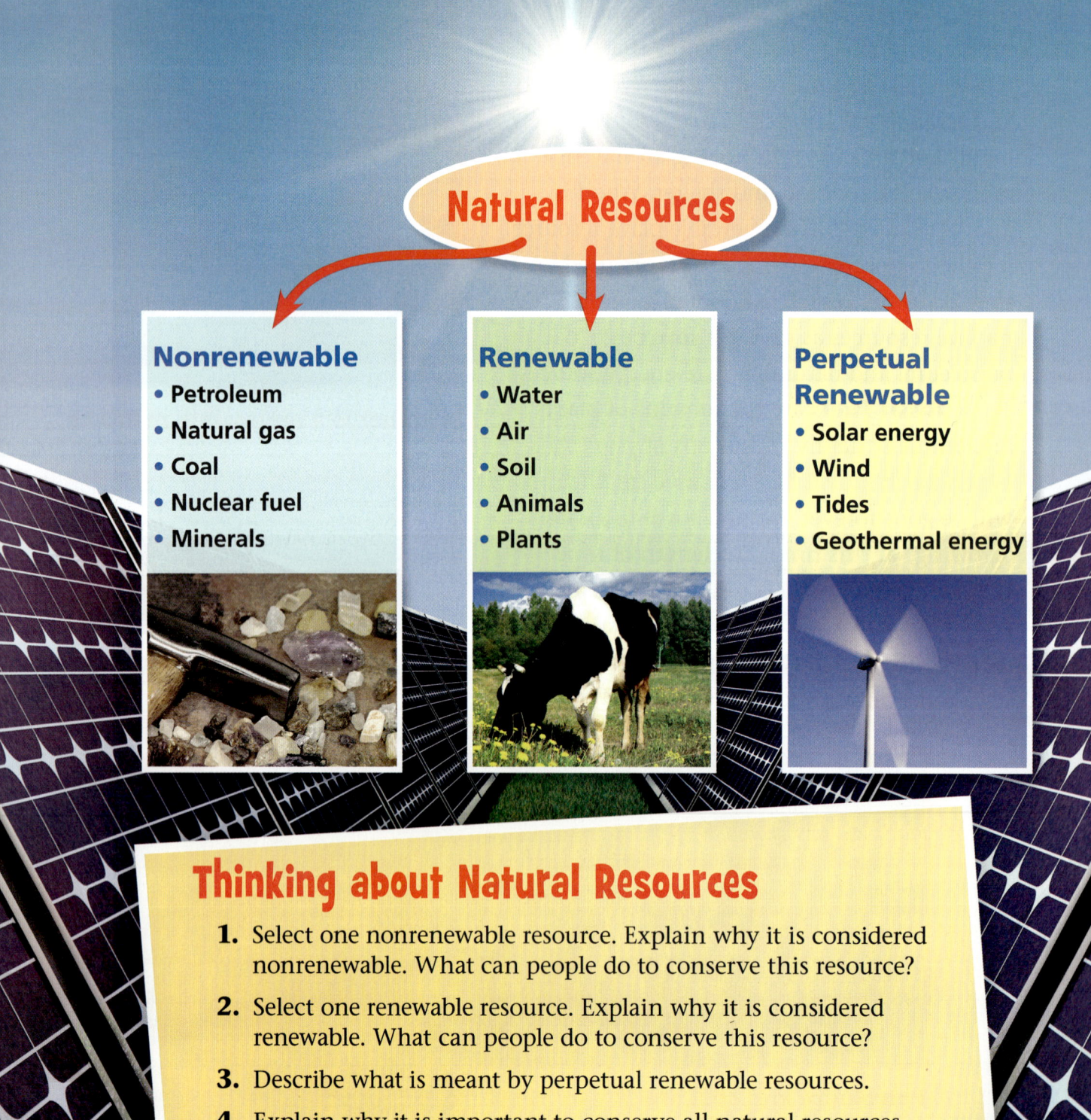

Natural Resources

Nonrenewable
- Petroleum
- Natural gas
- Coal
- Nuclear fuel
- Minerals

Renewable
- Water
- Air
- Soil
- Animals
- Plants

Perpetual Renewable
- Solar energy
- Wind
- Tides
- Geothermal energy

Thinking about Natural Resources

1. Select one nonrenewable resource. Explain why it is considered nonrenewable. What can people do to conserve this resource?
2. Select one renewable resource. Explain why it is considered renewable. What can people do to conserve this resource?
3. Describe what is meant by perpetual renewable resources.
4. Explain why it is important to conserve all natural resources.

The Power of Water

The water cycle moves water all over Earth. Energy from the Sun evaporates water and lifts it high in the air. The water condenses into clouds. Wind moves clouds all over Earth. Eventually the water falls from the clouds as rain, snow, sleet, or hail.

A lot of water falls high in the mountains. Water is matter. We know what happens to matter on a **slope**. The **force** of gravity moves it downhill. When water runs into something, it applies a force. Moving water has the force to push things around.

During very heavy rainstorms, rivers and streams can flood and overflow their banks. The force of the floodwater can wash away rocks and soil, destroy roads, and carry away cars and houses. The faster water flows, the more force it has, and the more damage it can do.

Hurricanes are strong storms that produce extremely high winds. When hurricanes come on land, they can cause a **storm surge**. A storm surge is a huge wall of water that washes onshore. On August 29, 2005, Hurricane Katrina hit New Orleans with a huge surge. The force of the surge plus the flow of the Mississippi River broke through the levees protecting the city. The resulting flood caused huge devastation. More than 1,800 people died, and the estimated cost of the damage was more than $100 billion.

A flood following heavy rain washed out a road.

The flood following Hurricane Katrina did massive damage.

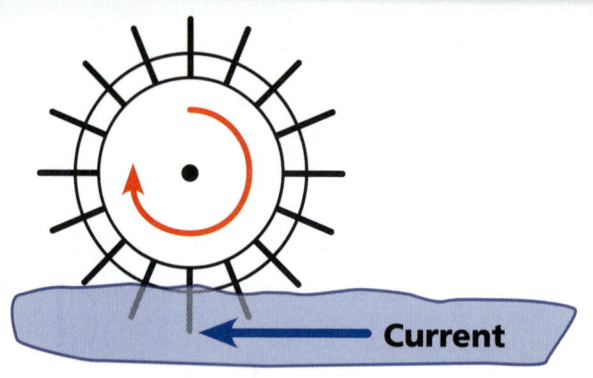

Using Water to Do Work

Were you able to put water to work? **Waterwheels** have been used for thousands of years. The early Greeks and Romans used them to grind corn. Early American towns used waterwheels to power gristmills and sawmills.

Two different forces can push or pull on a waterwheel. Moving water makes it turn. The force from the stream of fast-moving water hitting the **blades** pushes the **shaft** around. In this old-fashioned waterwheel, the current in the stream pushes on the blades at the bottom of the wheel to turn it.

Another way to drive a waterwheel is to fill "buckets" attached to the outside of a wheel. Water pours onto the top of the wheel. The buckets catch the water. The weight of the water pulling down turns the wheel. The water in the buckets spills out as the wheel turns.

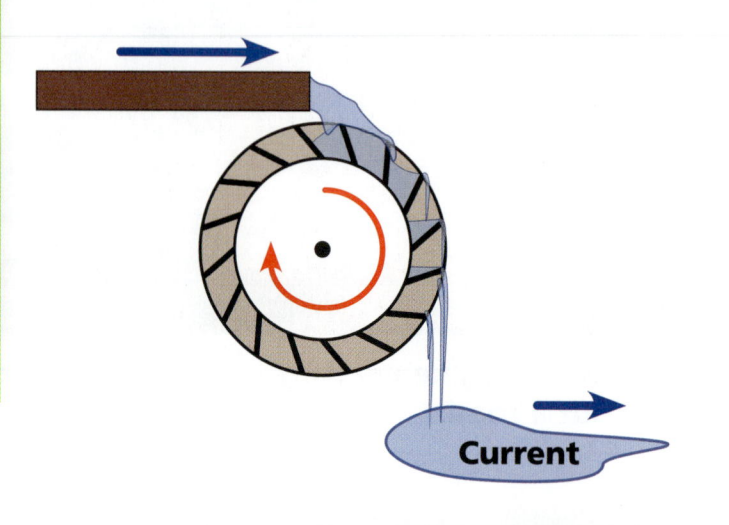

36

Huge generators are built into the base of Hoover Dam in Arizona and Nevada.

A modern kind of waterwheel is the **water turbine**. Water turbines are built into the bottoms of dams. Water from the reservoir behind the dam turns the turbine to generate **electricity**.

High-pressure water flows into a chamber above the turbine. Water then flows through the turbine, pushing on the turbine blades as it goes. The blades turn a shaft, which is connected to the generator.

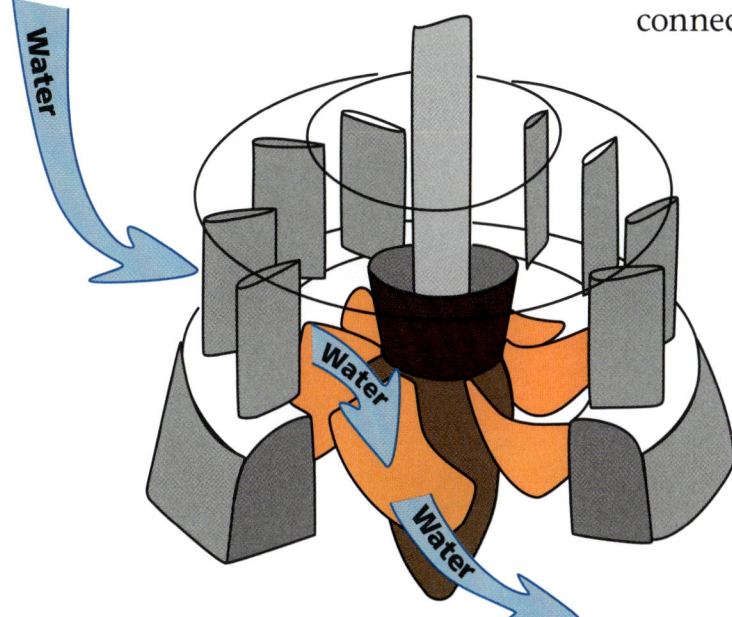

Water pushes the blades as it flows through the turbine.

Ellen Swallow Richards: An Early Ecologist

An American in 1900 could expect to live only to age 47. Today life expectancy is much longer. We owe that in part to Ellen Swallow Richards. She lived in a time when people understood little about germs and pollution. Yet Richards knew there was a connection between health and the environment. In the early 1900s, she wrote to the president of the Massachusetts Institute of Technology (MIT), "One of the most serious problems of civilization is clean water and clean air, not only for ourselves but for the world."

Ellen Swallow was born on December 3, 1842. She lived in Dunstable, Massachusetts. Growing up, Ellen did chores on her family's farm and helped in their store. She also took care of her mother, who was often sick. Ellen's first teachers were her parents. They saw that Ellen loved to learn. Before long, the family moved to Westford, Massachusetts, where Ellen entered school.

Ellen became a teacher after graduation. When her mother became ill again, Ellen returned home to help. But she was unhappy working in the family store. She wanted to learn more, and she wanted to go to college.

Richards in her study

Few colleges accepted women at that time. Many people believed studying hard would make women ill! But Ellen would not forget her dream. She worked at many jobs and saved all the money she could. Finally she had enough money to enter Vassar College. Vassar was an experimental school. It aimed to give women the same chance that men had to get an education.

Ellen was called a "special student" at Vassar because she was 26 years old. The other women were 14 to 19 years old. Ellen was too happy to care. Her favorite subjects were astronomy and chemistry. In 1870, she was part of Vassar's first graduating class.

Ellen planned to teach in Argentina, but war broke out. Instead she entered graduate school at MIT. She was not charged tuition. Ellen believed this was because she was poor. In fact, MIT was afraid to admit women. By not charging Ellen, the school could claim she was not really a student.

Ellen worked at MIT after her graduation in 1873. The professors respected her. One laboratory head said, "When we are in doubt about anything, we always go to Miss Swallow." Ellen married chemistry professor Robert H. Richards in 1875. They helped each other with their work.

Richards collecting water samples

In 1884, Ellen Swallow Richards became an instructor of "sanitary chemistry." For 2 years, she and Professor Thomas M. Drown studied the state's water supply. They suspected that something in the water was making people sick. Richards worked to find a way to test the **water quality**. Water was collected from every river and lake in Massachusetts once a month. Richards analyzed most of the 40,000 samples herself. When the survey was done, Massachusetts had the first standards for water purity. Professor Drown wrote that this was "mainly due to Mrs. Richards's great zeal and vigilance." From then on, Richards taught others how to analyze air, water, and sewage.

Ellen Swallow Richards started the Women's Laboratory at MIT in 1876. She wanted other women to study science. When the Women's Laboratory closed in 1883, Richards was thrilled. Through her efforts, women were no longer "special" at MIT. They were regular students, equal to men.

Ellen Richards with female students in 1888

Another of Richards's interests was nutrition. She opened the New England Kitchen, where immigrants were taught how to cook nutritious, inexpensive food. She cared deeply about public health. She urged women to eat right and exercise.

Ellen Swallow Richards died at the age of 68 on March 30, 1911. Many people consider her to be the founder of ecology. She said, "The quality of life depends on the ability of society to teach its members how to live in harmony with their environment." It was her belief that science should make people healthier. She worked hard to make that happen.

The MIT chemistry department in 1900

Solar Disinfection System

What happens if you are outside playing at recess and you get thirsty? You walk over to a drinking fountain for a drink. And if your dad needs to boil some water to make dinner, he goes to the sink and turns on the faucet.

In the United States, we don't usually think about how easy it is to get water that's safe to drink. We just turn on the faucet, and out comes clean water. Our water is treated and tested for safety before it gets to us.

What about other countries? Many people don't have running water in their homes and schools. They go to lakes and rivers to get water. Sometimes the water contains bacteria. Most bacteria don't hurt people, but some can make people sick. Boiling kills bad bacteria, but not everyone can boil their water whenever they need to.

In 1991, researchers in Switzerland set out to see if they could use everyday tools to make drinking water safe. They wanted to find a way to get rid of bad bacteria without expensive water treatment. The solution was a solar disinfection system, or SODIS. For SODIS, all you need is a clear plastic bottle and sunshine. SODIS works best in countries near the equator. That's where sunshine is strongest.

Here's how SODIS works.

1. Get a clean, clear plastic bottle with a cap. (Glass can be used, but plastic is best.)
2. Fill the bottle with water and put the cap on.
3. Lay the bottle flat on a piece of corrugated tin or on a roof.
4. Let the bottle lie in the sunshine for 6 hours to 2 days, depending on how cloudy it is. Then the water is ready to drink.

Light from the Sun is called **solar radiation**. Solar radiation heats the water and makes it safe to drink. Heat and ultraviolet radiation from the Sun, called UV-A, destroy the bad bacteria.

What's so great about SODIS? It costs nothing at all, and it recycles plastic bottles. Sometimes the simplest solutions are the best. Sunlight + water + a bottle = safe water. It's amazing!

1 Clean the bottles.

2 Fill the bottles with water.

3 Put the bottles in sunshine.

4 Wait 6 hours to 2 days.

5 Drink the safe water.

43

Science Safety Rules

1. Listen carefully to your teacher's instructions. Follow all directions. Ask questions if you don't know what to do.

2. Tell your teacher if you have any allergies.

3. Never put any materials in your mouth. Do not taste anything unless your teacher tells you to do so.

4. Never smell any unknown material. If your teacher tells you to smell something, wave your hand over the material to bring the smell toward your nose.

5. Do not touch your face, mouth, ears, eyes, or nose while working with chemicals, plants, or animals.

6. Always protect your eyes. Wear safety goggles when necessary. Tell your teacher if you wear contact lenses.

7. Always wash your hands with soap and warm water after handling chemicals, plants, or animals.

8. Never mix any chemicals unless your teacher tells you to do so.

9. Report all spills, accidents, and injuries to your teacher.

10. Treat animals with respect, caution, and consideration.

11. Clean up your work space after each investigation.

12. Act responsibly during all science activities.

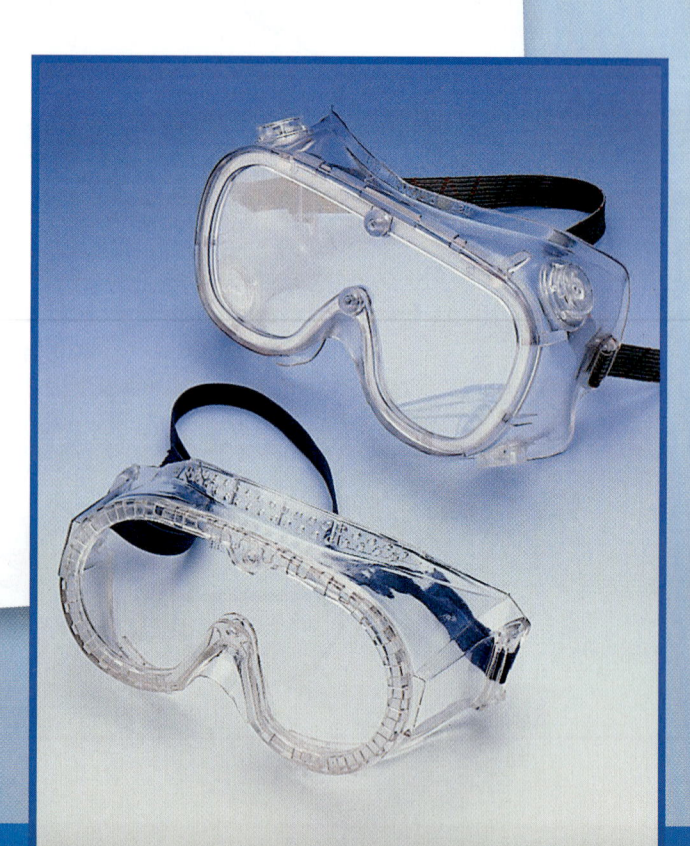

Glossary

absorb when a liquid soaks into a material

aquifer water that is underground in layers of rock or sediment

bead a dome-shaped drop of water

blade the part of a waterwheel that the water pushes as it moves downward

condensation the process by which water vapor changes into liquid water, usually on a surface

condense when water vapor touches a cool surface and becomes liquid water

conserve to use carefully and protect

contract to get smaller; to take up less space

decomposing organic matter humus; dead or discarded parts of plants and animals

density the amount of mass compared to the volume

dew water that condenses on a surface when the temperature drops at night

electricity energy that flows through circuits and can produce light, heat, motion, and sound

energy source a place where energy comes from, such as coal, petroleum, and natural gas

evaporate when liquid water in a material dries up and goes into the air

evaporation the process by which liquid water changes into water vapor

expand to get bigger; to take up more space

expansion an increase in volume

float to stay on the surface of water as a result of being less dense than water

force strength or power exerted on an object

fossil fuel the preserved remains of plants and animals that lived long ago and changed into oil, coal, and natural gas

freeze to change from a liquid to a solid state as a result of cooling

fresh water water that is in lakes, rivers, ground water, soil, and the atmosphere

gas a state of matter with no definite shape or volume; usually invisible

glacier a large mass of ice moving slowly over land

gravity the natural force that pulls objects toward each other. On Earth, all objects are pulled toward the center of Earth.

humus bits of dead plant and animal parts in the soil

hurricane a severe tropical storm that produces high winds

ice the solid state of water

iceberg a large mass of ice that has broken from a glacier and floats in the ocean

liquid a state of matter with no definite shape but a definite volume

mass the amount of material in something

matter anything that has mass

melt to change from a solid to a liquid state as a result of warming

mixture two or more substances together

more dense when an object has more mass for its size than another object. When an object sinks in water, it is more dense than water.

natural resource a material such as soil or water that comes from the natural environment

nonrenewable resource a natural resource that cannot be replaced if it is used up. Coal, petroleum, and natural gas are nonrenewable resources.

perpetual renewable resource a renewable resource that lasts forever. The Sun, wind, and tides are perpetual renewable resources.

precipitation rain, snow, sleet, or hail that falls to the ground

property something that you can observe about an object or a material. Size, color, shape, texture, and smell are properties.

renewable resource a natural resource that can replace or replenish itself naturally over time. Air, plants, water, and animals are renewable resources.

reservoir a place where water is collected and stored

retain to hold or continue to hold

runoff rain that does not evaporate or soak into the ground

salt water ocean water

shaft the part of a waterwheel that the blades turn

sink to go under water as a result of being more dense than water

slope a slanted or tilted surface

soak to be absorbed or move into another material

soil a mixture of humus, sand, silt, clay, gravel, or pebbles

solar radiation light from the Sun

solid a state of matter that has a definite shape and volume

storm surge when water piles up along a coast, rushing toward land faster than it can return to sea

surface area the area of liquid exposed to or touching the air

surface tension the skinlike surface on water (and other liquids) that pulls it together into the smallest possible volume

texture the feel or general appearance of an object or a material

thermometer a tool used to measure temperature

volume three-dimensional space

water a liquid earth material made of hydrogen and oxygen

water cycle the repeating sequence of condensation and evaporation of water on Earth, causing clouds and rain and other forms of precipitation

water quality a term used to describe the purity of water

water turbine a modern waterwheel

water vapor the gaseous state of water

waterwheel a wheel turned by the force of moving water

Index

A
Absorb, 19–22, 27, 45
Air, 5, 20, 23–24
Aquifer, 27, 45

B
Bead, 7, 45
Blade, 36, 37, 45

C
Cloud, 24–25
Condensation, 23–25, 45
Condense, 23, 45
Conserve, 30, 33, 45
Contract, 10, 17, 45
Current, 36

D
Dam, 9, 37
Decomposing organic matter, 31, 45
Density, 12, 18, 45
Dew, 23, 45

E
Ecologist, 38–41
Electricity, 37, 45
Energy, creating, 26–37
Energy source, 23, 24, 33, 45
Evaporate, 20, 23–25, 29, 35, 45
Evaporation, 19–20, 23–25, 29, 30, 45
    experiment, 21–22
Expand, 10, 17, 45
Expansion, 10, 45
Experiment, 21–22

F
Float, 11–12, 18, 45
Flow of water, 8–9
Force, 35–37, 45
Fossil fuel, 33, 45
Freeze, 17–18, 24, 26, 45
Fresh water, 4, 5, 14, 26, 45

G
Gas, 5, 20, 23, 45
Glacier, 5, 14, 24, 45
Graph, 22
Graph paper, 21
Gravity, 9, 35, 45
Ground water, 24, 25

H
Hail, 24. See also Precipitation
Humus, 31, 46
Hurricane, 35, 46

I
Ice, 4, 5, 13–18, 24, 26, 46
Iceberg, 15, 18, 46

L
Lake, 5, 28, 29
Liquid, 4, 5, 18, 20, 23–24, 46

M
Mass, 11–12, 18, 46
Matter, 9, 35, 46
Measuring tools, 21
Melt, 13, 46
Meter tape, 21
Mixture, 11, 31, 32, 46
More dense, 12, 18, 46
Mountain, 8–9, 14

N
Natural resource, 26, 31–34, 46
Nonrenewable resource, 33–34, 46

O
Ocean, 3, 4, 9, 28. See also Salt water

P
Particle, 10
Perpetual renewable resource, 34, 46
Pipes, 17–18
Plants, 24, 33
Population, 26
Precipitation, 24–25, 46
Property, 32, 46
Purify, 27, 30, 42–43

R
Rain, 24. See also Precipitation
Recycle, 24, 26
Refrigeration, 16
Renewable resource, 26, 32, 33, 46
Reservoir, 24, 37, 46
Results, 22
Retain, 32, 46
Richards, Ellen Swallow, 38–41
River, 5, 9, 24, 28, 29
Runoff, 29, 46

S
Safety rules, 44
Salt water, 3, 4, 26, 46. See also Ocean
Shaft, 36, 46
Sink, 11–12, 46
Sleet, 24. See also Precipitation
Slope, 35, 46
Snow, 24. See also Precipitation
Soak, 8, 46
SODIS. See Solar disinfection system
Soil, 8, 24, 31, 46
    properties of, 32
Solar disinfection system (SODIS), 42–43
Solar radiation, 43, 47
Solid, 4, 47
Storm surge, 35, 47
Sun, 23, 24
Surface area, 21–22, 47
Surface tension, 6–7, 47

T
Temperature, 17
Texture, 32, 47
Thermometer, 10, 47
T-table, 22

V
Volume, 7, 10, 12, 18, 47

W
Waste water, 28
Water
    as natural resource, 31–34
    as renewable resource, 33
    as vital resource, 26
    changes in, 10, 19–22
    defined, 47
    density, 12
    Earth's surface cover, 3–5
    flow of, 8–9
    in our cities, 29
    in our homes, 27
    leaving our homes, 28
    power of, 35–37
    purifying, 42–43
    states of, 4–5, 13–18
    uses for, 4, 36–37
Water conservation, 30
Water cycle
    defined, 23, 47
    description of, 23–25, 35
    water renewal, 26, 33
Water quality, 27, 40, 47
Water strider, 6, 7
Water turbine, 37, 47
Water vapor, 5, 19–20, 23–25, 47
Waterwheel, 36, 47

48